NEW VANGUARD 222

SELF-PROPELLED ANTI-AIRCRAFT GUNS OF THE SOVIET UNION

MIKE GUARDIA   ILLUSTRATED BY HENRY MORSHEAD

First published in Great Britain in 2015 by Osprey Publishing,
PO Box 883, Oxford, OX1 9PL, UK
PO Box 3985, New York, NY 10185-3985, USA
E-mail: info@ospreypublishing.com

Osprey Publishing, part of Bloomsbury Publishing Plc

A CIP catalog record for this book is available from the British Library

Print ISBN: 978 1 4728 0622 2
PDF ebook ISBN: 978 1 4728 0623 9
ePub ebook ISBN: 978 1 4728 0624 6

Index by Alan Rutter
Typeset in Sabon and Myriad Pro
Originated by PDQ Media, Bungay, UK
Printed in China through Worldprint Ltd

16 17 18 19   10 9 8 7 6 5 4 3 2

Osprey Publishing is supporting the Woodland Trust, the UK's leading
woodland conservation charity, by funding the dedication of trees.

www.ospreypublishing.com

Title page image: A Hungarian ZSU-57-2 (Photo courtesy VargaA/CC-
BY-SA-4.0)

CONTENTS

INTRODUCTION   4

ZSU-37 AND THE EARLY YEARS OF SOVIET AIR DEFENSE   6

ZSU 57-2   9

ZSU 23-4   18

9K22 TUNGUSKA (SA-19 GRISON)   39

TRENDS IN POST-SOVIET SPAAGS   45

FURTHER READING   47

INDEX   48

SELF-PROPELLED ANTI-AIRCRAFT GUNS OF THE SOVIET UNION

INTRODUCTION

During the Cold War, the size and power of the Red Army's armored and mechanized forces struck fear into strategists in the West. While historians and military analysts now agree that much of Soviet equipment was inferior to NATO's, Soviet mechanized forces were able to accomplish something their US counterparts never did: the development of a tracked, radar-guided, self-propelled antiaircraft gun (SPAAG) that could keep pace with the heavy armored formations.

Self-Propelled Antiaircraft Guns of the Soviet Union examines four SPAAGs from the Soviet era: the ZSU-37, ZSU 57-2, ZSU 23-4, and the 9K22 Tunguska (also known by its NATO reporting name, SA-19 Grison).[1] The ZSU-37 was the first series-produced SPAAG developed for the Soviet military. Although built to counter the Luftwaffe on the Eastern Front, its late arrival in 1945 precluded it from seeing any action in World War II. The ZSU 57-2 was the first Soviet SPAAG to see a wide export market. In Southeast Asia, the ZSU 57-2 served with the People's Army of Vietnam (PAVN) during the Easter Offensive of 1972, while in the Middle East, the Syrians and Egyptians used it as a multi-purpose gun platform during the Six Day War and the Yom Kippur offensives. The ZSU 23-4 was considered a grave threat to NATO aircraft. Many US Department of Defense publications were dedicated to examining how to defeat the ZSU 23-4 and its radar tracking system. In the late 1980s, the introduction of the 9K22 pioneered the concept of combining antiaircraft guns and surface-to-air missiles on a single, tracked platform. Although the Soviet Union dissolved in 1991, a few years after the 9K22's introduction, the Tunguska has remained in service with the Russian Ground Forces and saw limited action in the 2008 South Ossetia War.

During and after the Cold War, these Soviet-era SPAAGs often found roles beyond the realm of traditional air defense. In Afghanistan and Chechnya, the ZSU 23-4 accompanied armor formations into battle. The ZSU 23-4's guns could elevate to engage insurgents who stood on rooftops and mountainsides – areas beyond the elevation range of a T-72 or BMP main gun. During Operation *Desert Storm*, frontline Iraqi units were equipped with several ZSU 23-4s. Some US commanders considered the ZSU such a threat that they ordered their tank crewmen to engage it before engaging Iraqi tanks.

1   The 9K22 was named the "Tunguska" after the river in Siberia.

Although some Western militaries developed SPAAGs comparable to the ZSU series, such as West Germany's *Flakpanzer Gephard*, the United States frequently lagged behind the Warsaw Pact when it came to developing tactical air defense vehicles. During the Korean War, for example, the US Army developed the M42 Duster – a double-barreled, 40mm antiaircraft gun mounted atop an M41 tank chassis. By 1959, however, Army planners decided that SPAAGs had become obsolete in an era of high-performance jet aircraft. The Army therefore dumped its remaining M42s on the National Guard. Ironically, by the beginning of the Vietnam War, the Army realized the need for a low-altitude air defense gun and quickly recalled the M42 to active service. Nevertheless, the M42 ultimately found more use as a convoy and perimeter security vehicle than as a pure air defense weapon. Meanwhile, the Soviet Union stepped up production of the ZSU 57-2 and began laying the groundwork for the ZSU 23-4.

After the unveiling of the ZSU 23-4, the United States fielded the M163 Vulcan Air Defense System (VADS). The M163 combined the existing Vulcan antiaircraft gun with the M113 Armored Personnel Carrier. The M163 was a marked improvement over the Duster, as it had an optical fire control system and a range-finding radar. However, the VADS's firing range was too short to justify installing a target-tracking radar. By the early 1980s, however, the Army had already developed plans for a larger, radar-guided SPAAG that could fight alongside the emerging M1 Abrams Main Battle Tank and M2 Bradley Fighting Vehicle. The result was the experimental M247 Sergeant

In the early days of the Cold War, the M42 Duster was NATO's only tracked air defense gun. Like its Soviet counterpart, the ZSU 57-2, the M42 was a "fair-weather" air defense system that could only engage low-performance aircraft. While the Soviet Union increased production of the ZSU 57-2, the United States withdrew the M42 from its frontline formations in the late 1950s – declaring the SPAAG obsolete in an age of jet aircraft. At the beginning of the Vietnam War, the Duster was recalled to active service, serving mostly as a convoy security vehicle. (US Army)

York Division Air Defense (DIVAD) vehicle. Built atop an M48 tank chassis, the M247 carried twin 40mm guns and, like the ZSU 23-4, also had a tracking radar system. However, soaring costs, developmental problems, and troublesome field testing prompted Defense Secretary Caspar Weinberger to cancel the program in 1985.

Following the breakup of the Soviet Union, many of the SPAAGs within the Soviet Army were absorbed by the various successor states, while others were sold to allied nations. Of the four SPAAGs covered in this book, only the Tunguska remains in active production.

ZSU-37 AND THE EARLY YEARS OF SOVIET AIR DEFENSE

Following the Russo–Finnish War of 1940, the Red Army rebranded its existing Air Defense Directorate into the Main Administration of the Air Defense Force (GU-PVO). However, the GU-PVO only took responsibility for training and equipping the air defense units. Command and operational control of these units remained within the Soviet Military Districts. This reorganization, however, proved to be of little consequence; the Germans had already begun conducting aerial reconnaissance missions over the Russian borderlands to prepare for Operation *Barbarossa* – Hitler's bold invasion of the Soviet Union in the following year. On August 14, 1939, days before the start of World War II, the German and Soviet foreign ministers, Ribbentrop and Molotov, had signed a non-aggression pact. Hitler, however, had no intention of abiding by the agreement and ordered the reconnaissance flights to identify weaknesses along the western frontier of the Soviet Union. Thus, by the time the Germans invaded in June 1941, the Luftwaffe was able to destroy much of the Soviet Air Force when it was still on the ground.

 ZSU-37
The ZSU-37 was fielded to Soviet forces in 1945. By this time, however, the Luftwaffe had all but disappeared from the skies over Russia. The inaugural ZSU, therefore, saw no combat along the Eastern Front. Featuring a single-barreled 37mm gun, the ZSU-37 stood atop an SU-76M chassis and was powered by GAZ-203 engine. Only 75 units were built before production ended in 1948.

In 1945, Soviet engineers had developed an air defense vehicle featuring a single 37mm gun mounted atop an SU-76 chassis. Named the *Zenitnaya Samokhodnaya Ustanovka*, or ZSU for short, the weapon took the suffix of its gun caliber, thus becoming the ZSU-37. Although fielded in the final months of World War II, the ZSU-37 did not see any combat along the Eastern Front, as the Luftwaffe had largely disappeared from the skies over the Soviet Union. (Russian Defense Ministry)

The Soviet Union, meanwhile, mounted a piecemeal air defense campaign that gradually became more effective as the war continued. To keep the Luftwaffe at bay, the Red Army employed a number of antiaircraft guns along the Eastern Front. These included the 37mm M1939 and the 40mm M1933, which had ranges of 3,660m and 3,050m respectively. For higher altitude engagements, the Soviets relied on various 76mm, 85mm, and 105mm pieces. Weapons of the latter caliber could often reach targets in excess of 10,000m.

In the midst of defending Mother Russia from the Nazi onslaught, Soviet engineers began experimenting with designs for a tracked, self-propelled air defense gun. The first of these designs was the T-90 SPAAG, a modified version of the T-70 light tank, featuring twin-barreled 12.7mm DshK machine guns with optical sighting. However, since the DshK had an effective range of only 2,000m, it was clear the Red Army needed a better design. Thus, the T-90 was scrapped in favor of the ZSU-37.

Sharing a chassis with the SU-76M field gun, the new SPAAG took the title *Zenitnaya Samokhodnaya Ustanovka* (Antiaircraft, Self-Propelled Mount), or "ZSU" for short. The primary armament consisted of a singular 37mm M1939 antiaircraft gun (hence the name ZSU-37). This mounted M1939 fired a total of 320 armor-piercing, incendiary, and fragmentation rounds. The vehicle carried a crew of six – a driver, loader, two aimers (one who controlled the gun's azimuth, one the elevation), and two sight adjusters (one to identify target speed and range, the other for tracking the target's dive angle and course).

Aboard the ZSU-37, the crewmen had access to some of the most high-tech instrumentation available to the Red Army. To facilitate target acquisition, the ZSU-37 featured an automatic sight with two collimators and a stereo range finder. The vehicle's speed and endurance were comparable to the SU-76M; made possible by the installation of a GAZ-203 engine. The ZSU-37 went into full-scale production in March 1945. By this time, however, the threat from the Luftwaffe had disappeared from the skies over the Soviet Union, leaving the ZSU-37 with virtually no role to play on the Eastern Front. Only 75 vehicles were built by the time production ended in 1948.

ZSU 57-2

By the beginning of the Cold War, it was clear that the Soviet Union needed a better air defense gun, as the ZSU-37 had a small chassis, low firing range, and only one barrel. But even while the ZSU-37 remained in active service, Soviet defense planners had begun laying the groundwork for a more resilient SPAAG. In February 1946, designers at Works No. 174 in Omsk, and at Research Institute No. 58 in Kaliningrad, submitted a joint proposal for a tracked SPAAG based on the T-34 tank chassis. However, this design was scrapped in favor of Vasily Grabin's idea to build a double-barreled SPAAG based on the emerging T-54 chassis. Grabin was a Soviet artillery designer and the Chief of Designs at the Joseph Stalin Factory No. 92 in Gorky. During the Second World War, Factory No. 92 produced some of the finest artillery pieces in the history of the Red Army. In fact, the ZiS-3 field gun (Grabin's own design) was the most-produced Soviet cannon of World War II. Grabin's new SPAAG design called for the development of a twin-barreled 57mm auto cannon capable of firing more than 200 rounds per minute. Grabin's final product, known as the Ob'yekt 500, was completed in 1948 and entered production as a prototype in the summer of 1950.

Mounted atop the T-54 chassis, the Soviet Army called the Ob'yekt 500 the ZSU 57-2. Retaining the *Zenitnaya Samokhodnaya Ustanovka* designation, this new SPAAG's suffix "57-2" referred to the twin-mounted 57mm cannons. Initial field testing took place in the spring of 1951, which included mobility trials and a 2,000-round test fire with the main guns. Following these initial field stakes, six more prototypes were built. Each successive prototype offered upgrades in ammunition stowage, among other improvements. The final field tests were completed in 1954 and the ZSU 57-2 officially entered service on February 14, 1955.

The ZSU 57-2 on parade in Red Square, November 7, 1964. After the ZSU-37 was retired in the late 1940s, Soviet designers began work on a more resilient and larger caliber air defense vehicle. Mounted atop a modified T-54 chassis, and featuring twin 57mm auto cannons, the ZSU 57-2 entered service with the Red Army on February 14, 1955. (ITAR-TASS/ Sovfoto)

A Polish ZSU 57-2 on one of its many parades through Warsaw. The ZSU 57-2 was the first Soviet SPAAG to see a wide export market. Members of the Warsaw Pact and other communist allies around the world received sizeable quantities of the open-turret SPAAG from the Soviet Union. (Polish Defense Ministry)

The ZSU 57-2 carried a crew of six: the commander, driver, gunner, loader, and two sight adjusters. The T-54 chassis proved to be a sturdy and reliable fit for the ZSU. However, to compensate for the weight of the oversized turret (and the SPAAG's greater mobility requirements), Soviet engineers modified the chassis to have four road wheels on either side of the vehicle and lighter armor than the standard Soviet battle tanks.

The ZSU's large open-topped turret was placed atop a ball-bearing ring that measured 1,850mm in diameter. The turret could traverse 360 degrees and was powered by a direct-current electric motor and hydraulic speed gears. When powered electrically, the ZSU's turret could traverse at a maximum rate of 36 degrees/sec. Protruding prominently from the turret were the twin air-cooled, recoil-operated, S-68 57mm auto cannons. The armament system weighed 4,500kg and recoiled more than 300mm when firing. The gun's range of elevation was between -5 degrees and +85 degrees and could be elevated or depressed at nearly 20 degrees/sec. The twin-barreled S-68 made the ZSU 57-2 the most powerful SPAAG of its day. It had an impressive rate of fire even by Western standards: up to 240 rounds per minute and with a muzzle velocity of 1,000m/sec. The gun fired both fragmentation and armor-piercing rounds. Each 57mm projectile weighed about 2.8kg and contained 1.2kg of nitro-cellulose powder. When fired horizontally, the projectile had a maximum effective range of 12km. When firing at the typical slant of an antiaircraft engagement, however, the effective

B | **ZSU 57-2**

Featuring twin-mounted 57mm cannons, the ZSU 57-2 entered service on February 14, 1955. The 57-2 was based on the T-54 chassis, but to make the vehicle more maneuverable, Soviet designers modified the design to have four road wheels instead of five. The vehicle carried a crew of six: the commander, driver, gunner, loader, and two sight adjusters. It was also the first SPAAG to see a wide export market. Members of the Warsaw Pact, and Communist allies around the world, received various quantities of the 57-2. Its service life, however, was largely cut short by the development of the ZSU 23-4. As the 57-2 was a fair-weather SPAAG, and had no tracking system, it days were numbered by the time the 23-4 entered service. Nonetheless, the ZSU 57-2 remains in service with a handful of nations as a dual-purpose air defense/ground support vehicle.

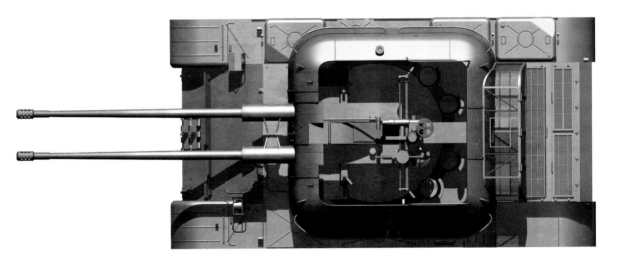

range fell to about 6km. The usual armor-piercing rounds aboard the ZSU 57-2 were capable of penetrating up to 110cm of armor. A total of 300 rounds were carried inside the ZSU: 176 clipped rounds loaded in the turret, 72 clipped rounds in the hull's front, and 52 separate, unclipped rounds kept in a special stowage underneath the turret floor.

The ZSU's armor protection, though lighter than a main battle tank, could nevertheless stop the normal variety of small arms fire. The probability of penetration for any larger caliber weapons, however, depended upon where the projectile hit. The ZSU's armor protection was only 15mm at its thickest, with the heaviest allocations given to the hull's front and bottom, the turret sides, and the gun mantle.

Propulsion was accomplished by a V-54 12-cylinder, 4-stroke diesel engine. Even with the ZSU's tare (unladen) weight of 28 metric tonnes, the V-54 produced enough output to give the vehicle 520 horsepower. This gave the ZSU 57-2 a top speed of 50km/h (31mph) on improved surfaces and 30km/h (18mph) when travelling off-road. Fuel was delivered to the V-54 engine by three fuel tanks – one internal, two external – with a combined capacity of 850l. Its operational range varied from 320km to 420km depending on terrain.

Still, for its size, the ZSU 57-2 was highly maneuverable. During field trials, the ZSU negotiated vertical obstacles nearly 1m high, crossed trenches measuring 3m across, and could ford water obstacles 1.4m in depth. The vehicle could also negotiate gradients of up to 30 percent before it became susceptible to rollover. Because the ZSU had a high power-to-weight ratio (18.6 horsepower per metric tonne) it had better acceleration than most of its armored stable mates.

Although the ZSU 57-2 was the most powerful SPAAG of its day, the vehicle had several drawbacks. First, its turret lacked an armored overhead cover for its crewmen. This left the crew dangerously susceptible to both aerial and ground fire. Second, it lacked a search, range-finding, or fire control radar. Because target acquisition was accomplished via optical reflex sights, the 57-2 could not engage targets at night or during periods of low

A long formation of ZSU 57-2s round a corner in Moscow's Red Square during the May Day Parade of 1962. The ZSU 57-2 was a regular sight during the Soviet military parades of the late 1950s and early 1960s. By 1970, however, the ZSU 57-2 had largely been replaced by newer air defense vehicles. The remaining 57-2s were either retired or given to allied nations, such as Vietnam. (ITAR-TASS/Sovfoto)

visibility. Third, the 57mm cannon's rate of fire was still not fast enough to ensure a high kill-per-burst probability. This was compounded by the fact that the ZSU's air-cooled barrels required more time to cool off after rapid-fire engagements. In fact, a battery of four ZSU 57-2s had a lower firing efficiency than a six-gun battery of towed S-60 antiaircraft pieces. Fourth, the turret's slow rate of traverse made it practically incapable of intercepting high-performance aircraft at low altitudes.

Soviet leaders quickly recognized the SPAAG's shortcomings and began a modernization program for it in 1957, even as the ZSU 57-2 was still in its initial production run. This modernization called for the installation of a radar system aboard the ZSU 57-2, but the plan was scrapped in favor of building another SPAAG atop a newer tank chassis. This became the ZSU 23-4.

Despite its flaws, the ZSU 57-2 was the Soviet Union's primary SPAAG for nearly a decade. It was also the first Soviet SPAAG to see a wide export

The People's Army of Vietnam made extensive use of the ZSU 57-2 during the latter years of the Vietnam War. Here, a knocked-out ZSU lies in ruin near An Loc, destroyed during the 1972 Easter Offensive. (Patton Museum)

market. Throughout the late 1950s and early 1960s, the closest members of the Warsaw Pact – such as East Germany (GDR), Poland, Hungary, and Romania – and other Soviet allies received limited quantities of the 57-2. East Germany, being the first foreign operator of the ZSU 57-2, received its vehicles in September 1957. Over the next four years, the East German *Nationale Volksarmee* received 129 units of the ZSU 57-2, but began replacing them with the ZSU 23-4 by the end of the 1960s. By 1979, the East Germans had removed the ZSU 57-2 from active service, although some were converted into utility vehicles for the purpose of training T-54 tank drivers. Following German reunification in 1990, these training vehicles were passed to the *Bundeswehr*, but were ultimately scrapped with the rest of the GDR's military equipment.

Poland received comparable shipments of the ZSU 57-2, the vehicle serving in the Polish Army until the 1970s. Like their comrades in East Germany, the Poles eventually phased out the ZSU 57-2 in favor of the newer 23-4. Romania, Hungary, and Bulgaria received the ZSU 57-2 as military hand-me-downs from other Warsaw Pact members after the ZSU 23-4 became available. Czechoslovakia imported one ZSU 57-2 for testing, but rejected it in favor of the Czech-built M53/59 Praga SPAAG. The former Yugoslavia, despite its political non-alignment with the Warsaw Pact, fielded more than 100 ZSU 57-2s between 1963 and 1964. Unlike their counterparts in the Soviet Bloc, however, Yugoslav forces kept the ZSU 57-2 in active service until well into the 1990s. After the breakup of Yugoslavia in 1992, the ZSUs were used by the various successor states, many of them seeing action in the Balkan Wars.

ZSU 57-2 IN VIETNAM

C

A North Vietnamese ZSU 57-2 fires at US warplanes along the Ho Chi Minh Trail during the Easter Offensive, 1972. The People's Army of North Vietnam made extensive use of the ZSU 57-2 during the latter years of the Vietnam War. Prior to their acquisition of the ZSU, however, North Vietnam's air defense forces had relied on the Chinese-built Type 63 SPAAG. The ZSU 57-2 remains in service with the unified Vietnamese Army, but it has largely been replaced by the ZSU 23-4.

Finland imported a dozen ZSU 57-2s in the early 1960s, where they remained in service over the next several decades. The Finnish Army recently undertook a program to modernize its fleet of ZSUs, which included provisions for adding a radar system, but the program was cancelled due to budgetary constraints. The Finns retired their last ZSU 57-2 in 2006. Cuba also imported the ZSU 57-2, and although it later took delivery of the ZSU 23-4, the original 57-2s remain in service to this day.

In the Middle East, Egypt and Syria were the largest operators of the ZSU 57-2. Egypt ordered 100 units from the Soviet Union in 1960. Decades later, Egyptian defense planners initiated a modernization program for the 57-2, which equipped the SPAAG with radar. As of 2010, at least 40 ZSUs remained in service with the Egyptian Army. Syria placed orders for more than 200 ZSU 57-2s in 1966, most of which were delivered between 1967 and 1973. Although the Syrian ZSU 57-2 eventually lost favor to the newer ZSU 23-4, a few 57-2s remained in service as of 2014.

Iran and Iraq each ordered 100 ZSUs from the Soviet Union during the 1960s. The Iraqis used the ZSU 57-2 throughout the 1980s and briefly into the 1990s. Like their Arab neighbors, Egypt and Syria, Iraq slowly replaced the ZSU 57-2 with the ZSU 23-4. In Iran, the 57-2 saw action on the frontlines of the Iran–Iraq War, but its service history thereafter remains largely unknown. Around 90 vehicles remained in service with the Iranian Army until 2002.

In the Far East, North Vietnam ordered nearly 100 ZSUs during its war with the United States and South Vietnam. At its peak, the PAVN had 500 ZSU 57-2s in service. After the fall of Saigon in 1975, all ZSU 57-2s were absorbed into the new army of a unified Vietnam. Meanwhile, North Korea imported around 250 ZSU 57-2 turrets for installation aboard the T-59 tank chassis to create its own SPAAG. These amalgamated vehicles are reportedly still in service.

At the time of writing, the ZSU 57-2 remains in service with only a handful of countries. Of the former Eastern Bloc, Bulgaria is the only country still using the vehicle. Egypt, Syria, and Algeria have retained some of their older ZSUs, while in sub-Saharan Africa, the armies of Angola, Mozambique, and Ethiopia still regularly use the vehicle.

The ZSU 57-2 first saw combat in the Arab–Israeli conflicts of the 1960s and 1970s. The Egyptians and Syrians had fully incorporated the ZSU into their formations before the start of the Six Day War in 1967. During the conflict, Israeli forces clashed with their Arab neighbors – Syria, Egypt, and Jordan – and won a decisive land war which wrested the Sinai Peninsula, the West Bank, and the Golan Heights from Arab control. In the opening stages of Israel's advance into Egypt, the Israeli 7th Armored Brigade (equipped with M48 Patton tanks) met a battery of Egyptian ZSU 57-2s guarding the perimeter of the El-Arish airstrip. Upon sighting the incoming tank formations, the Egyptian ZSU gunners leveled their 57mm cannons and fired in successive volleys. Against the M48s' 120mm armor, however, the antiaircraft shells did nothing to impede the Israelis' march towards the airfield. Although the ZSUs were reinforced by platoons of T-34 tanks, these Egyptian defenders proved no match against the firepower of the M48s.

For the People's Army of Vietnam (PAVN), the arrival of the ZSU 57-2 was a welcome relief. Although it arrived too late to have any real effect on the air defense campaign, the ZSU 57-2 was a vast improvement over the

An Arab ZSU 57-2 captured by the Israeli Defense Force stands on display at the Yad La-Shiryon Museum. Egypt and Syria were the most significant foreign recipients of the ZSU 57-2. Both countries used the 57-2 in their wars against Israel in 1967 and 1973. (Bukvoed RFI)

PAVN's Type 63. The Type 63 was a self-propelled air defense gun based on the Chinese Type 58 tank, which itself was a copy of the Soviet T-34. A twin-barreled 37mm auto cannon served as the primary armament, mounted atop an open turret. Because the Type 63's turret lacked any hydraulic systems, the vehicle was slow to engage aerial targets. Therefore, the PAVN made little use of the Type 63 as an air defense vehicle and – like its American counterpart, the M42 Duster – it saw more action as a ground support weapon. Although gradually phased out and replaced by the ZSU series, the Type 63 did see limited action during the Easter Offensive of 1972, where at least one vehicle was captured by South Vietnam's Army of the Republic of Vietnam (ARVN).

The ZSU 57-2, however, took the lead during the PAVN's air defense operations of the Easter Offensive. This 1972 campaign was a massive land invasion that was intended to destroy as many ARVN units as possible and strengthen North Vietnam's bargaining position at the ongoing Paris Peace Accords. The North Vietnamese had also timed the Easter Offensive to coincide with the monsoon season, the low 500ft (152m) cloud ceiling of which negated the use of any high-altitude airstrikes. Thus, if the Allied air forces wanted to disrupt the PAVN's movement, they would have to do so

LEFT
A Croatian ZSU 57-2, with its unorthodox camouflage scheme, sits at a war museum in Vukovar, Croatia. The ZSU 57-2 saw widespread service during the wars that followed the break-up of Yugoslavia. (Toca RFI)

RIGHT
A Polish ZSU 57-2 stands flanked by a larger ZSU 23-4 at the Museum of Polish Military Technology in Warsaw. After a decade in service, the Poles began replacing the obsolete 57-2 with the newer and better-performing 23-4. Other members of the Warsaw Pact soon followed. (Mateusz Kornecki)

with close air support, and expose themselves to the ZSU's line of fire. As 30,000 PAVN soldiers drove south through the DMZ, the ZSU batteries (together with various 85mm and 100mm towed pieces and dismounted soldiers carrying the shoulder-fired "Strela 2" missile) gave continual cover to the PAVN's forward bases and supply nodes. Although the ZSU 57-2 scored some hits during the Easter Offensive, it was easily outgunned and outmaneuvered by the F-4 Phantom and other Allied aircraft.

During the First Gulf War, the Iraqis used the ZSU 57-2 in limited quantities. By this time, they had been gradually withdrawn from frontline units to make way for the newer 23-4. Nevertheless, on January 16, 1991, one ZSU 57-2 succeeded in downing a British Tornado GR1 strike fighter during an RAF raid on the Shaibah air base. Later in the 1990s, the ZSU 57-2 saw combat yet again during the Balkan Wars in the former Yugoslavia. However, most of these engagements saw the ZSU being used as a ground assault weapon within the Serbians' mobile gun batteries. During the Croatian War of Independence, Croat forces succeeded in capturing two 57-2s from the Yugoslav People's Army. During the 1999 NATO bombing campaign, Serbian forces employed the ZSU 57-2 against incoming aircraft, but with negligible effect on the bombing campaign.

ZSU 23-4

By the late 1950s, Soviet defense planners had realized that the ZSU 57-2 had become obsolete. Thus, the push for a new SPAAG, built atop a newer tank chassis, with improved armaments, and a radar system began even while the ZSU 57-2 remained in serial production. On April 17, 1957, the Soviet

The ZSU 23-4, so named for its quadruple-barreled 23mm gun system, was born of the deficiencies found in the ZSU 57-2. The ZSU 23-4 featured a closed turret and a state-of-the-art radar system that could track its targets. Throughout the Warsaw Pact, it was nicknamed the "Shilka" – after the river on the Russia–China border. On the NATO side, it was frequently called "Zeus" – an informal pronunciation of its abbreviated name, ZSU. (US Army)

The ZSU 23-4 on parade in Red Square, November 7, 1967, celebrating 50 years since the Bolshevik Revolution. (ITAR-TASS/Sovfoto)

Council of Ministers adopted Resolution 426-11, which called for the simultaneous development of two all-weather, radar-guided air defense platforms. These parallel designs were the ZSU 37-2 (known as the Yensei) and the ZSU 23-4 (known as the Shilka).[2] The idea was that the Yensei would provide mobile air defense for armored formations while the Shilka would provide cover for motorized infantry units.

The ZSU 37-2 was developed at Experimental Design Bureau No. 16 in Moscow and featured twin 37mm auto cannons. Its chassis was similar to the SU 100 self-propelled field gun and its armor could withstand 7.62mm gunfire from ranges up to 400m. The Yensei's effective firing range was comparable to the ZSU 57-2. During field tests, Soviet evaluators determined that the guns could fire 150 rounds in a continuous burst, followed by a 30-second cool-off period to keep the barrel from overheating. The 37-2 could also reach a top ground speed of about 30km/h.

Meanwhile, the ZSU 23-4 was based on the chassis of the PT-76 light amphibious tank. Its armored skin was made of welded steel varying in thickness from 9.2mm in the turret to 15mm in the hull. This configuration gave the ZSU 23-4 reasonable protection from a variety of NATO small arms. Its power plant was an 8-cylinder, 6-stroke, 20-liter engine capable of producing 280 horsepower which could propel the vehicle to a top road speed of 50km/h and a maximum off-road speed of 30km/h. Despite its girth, the prototype ZSU 23-4 was highly maneuverable, with operational ranges of up to 450m. It could engage targets effectively while driving on inclines of up to 30 degrees and could ford water obstacles at depths of up to 1m. The vehicle itself weighed 19 metric tonnes and accommodated a crew of four: a driver, commander, gunner, and radar operator.

The prototypes for the ZSU 37-2 and ZSU 23-4 were completed in December 1960 with field tests continuing until October 1961. However,

2　Both the Yensei and Shilka were named after rivers in Siberia.

after these parallel field trials, the Council of Ministers determined that the ZSU 23-4's burst-to-kill ratio was nearly twice as high as the ZSU 37-2. Although the Yensei had a greater effective range and its manufacturing costs were comparable to Shilka, the Council adopted the Shilka on September 5, 1962 and further work on the Yensei was terminated. After two more years of refinement and recalibration, the first ZSU 23-4 was delivered to the Soviet Army in 1965. Although full-scale production ended in 1984, the vehicle remains in service with the Russian Ground Forces, Russian Marines, and the militaries of more than 30 countries.

The quad-mounted 23mm 2A7 guns are among the most distinctive features of the ZSU 23-4. These auto cannons fire a standard, steel-cased 23x152mm bullet at a muzzle velocity of 970m/sec. The singular 2A7 has a cyclical rate of fire of 850–1,000 rounds per minute. This gives the quad gun mount a combined rate of fire of 3,400–4,000 rounds per minute. The ZSU 23-4 typically carries two classes of ammunition: High-Explosive (HE) rounds and Armor-Piercing (AP) rounds, with tracer rounds available for each type. The ZSU's ammunition belt contains 40 HE and ten AP rounds. Fired horizontally, the 2A7 guns have a maximum range of 7km (4.3 miles). Their maximum vertical range is 5.1km (3.2 miles) with an effective range of 1 to 2.5km.

This gun battery gave the ZSU 23-4 more lethality than its closest American counterpart, the M163 Vulcan Air Defense System (VADS). As a baseline measurement, a ZSU 23-4 firing a 40-round burst against an F-4 (in level flight) from a range of 1km had a kill-per-burst probability of 0.13. The VADS, firing at a MiG-21 with a 60-round burst under the same conditions, had a kill-per-burst probability of only 0.08. By comparison, this gave the ZSU a greater effective range of 66 percent and made it almost 50 percent more accurate.

In field tests and in combat, the ZSU 23-4 performed most effectively against attack helicopters, but the kill-per-burst probability depended on three things: target range, the ZSU's exposure time, and maneuverability of the target. Naturally, hovering targets were easier to engage than those that maneuvered, but for either type of target, the ZSU's accuracy dropped off considerably beyond 2,500m. Still, at ranges of 1–3km, the Shilka had a higher kill-per-burst probability, so long as its exposure time did not exceed 50 seconds. During field tests, a helicopter versus Shilka engagement had the helicopter unmask from behind a hill, acquire the ZSU in its sights, fire its missile, and track the missile's trajectory to the target. Altogether, this process took the pilot 60–90 seconds to complete. However, the ZSU's battle drill, including target acquisition, firing, and rounds-to-target, took as little as 25 seconds.

D

ZSU 23-4

The ZSU 23-4 Shilka was the most popular SPAAG produced by the Soviet Union and, arguably, the most fearsome SPAAG of the Cold War. Featuring four 23mm guns and a revolutionary radar tracking system, the ZSU 23-4 caught the West by surprise. Several NATO publications were dedicated to showing pilots how to defeat or mitigate the threat from the ZSU's radar. Like its predecessor, the ZSU 57-2, the Shilka was exported to the Middle East and member nations of Warsaw Pact. Along the Iron Curtain and during the Gulf War, some commanders considered the Shilka such a grave threat that they ordered their tank crews to destroy it before engaging enemy tanks.

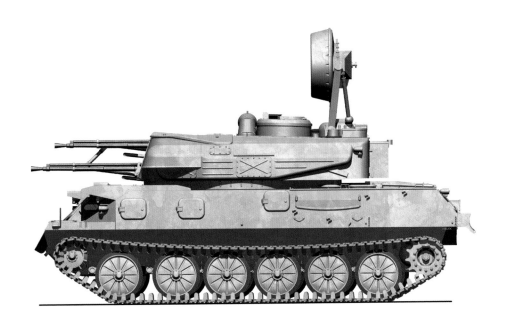

A pair of ZSU 23-4s leading a section of air defense vehicles during the annual Bolshevik Revolution parade. Trailing the 23-4s are the older ZSU 57-2s, which remained in Soviet service until well into the 1970s. (ITAR-TASS/Sovfoto)

Syrian ZSU 23-4s prepare for combat during the 1973 Yom Kippur War. After their defeat in the Six Day War, the Syrians and Egyptians solicited the Soviet Union for a better air defense platform than the ZSU 57-2. (US Army)

Thus, to mitigate the lethality of the ZSU, American aviators devised a series of tactical countermeasures. For instance, rotary pilots were advised to stay "unmasked" (rise above concealing terrain) for no more than 30 seconds. Because the weapon systems aboard NATO's attack helicopters had a greater range than the ZSU's, pilots were encouraged to engage the SPAAG at stand-off distances (3,000m and beyond). At these ranges, the AH-1 Cobra and AH-64 Apache could achieve an 80 percent kill probability. The ZSU, on the other hand, could achieve a kill probability of 10 percent at best.

Another technique employed by NATO pilots was to minimize their exposure height when engaging the ZSU. If operating from behind concealing terrain the pilots were advised to unmask at heights of no more than 10m beyond their concealment before engaging the SPAAG. Another technique was to have a scout helicopter identify the ZSU's position and simultaneously divert the SPAAG's attention while the attack helicopter zeroed-in for the kill. Alternatively, a ground-based scout could identify the ZSU and act as a forward air controller while guiding the helicopter to its target. Yet another method had the attack helicopter employed as a spotter for incoming artillery against the ZSU 23-4 and its parent formation. Ideally, this would kill the ZSU or disrupt it long enough for the helicopter to engage the vehicle at an appropriate stand-off range.

The fire control radar aboard the ZSU 23-4 (known as the RPK-2 or Tobol) can acquire targets at a line-of-sight range of up to 20km. The RPK-2 radar is a J-band device mounted atop two collapsible supports located towards the rear of the turret. It can effectively lock on to and track moving targets up to 10km away and can provide good range data for low-flying targets such as attack helicopters and fixed-wing, high-performance aircraft. For acquiring targets at lower altitudes, the radar is accompanied by a high-powered optics system providing 2x and 6x magnification. According to Soviet tactical doctrine, the optics became the primary means of target acquisition during periods of emission control or whenever the enemy was likely to use electronic countermeasures.

The RPK-2 radar put the ZSU 23-4 light years ahead of any comparable air defense system in NATO. Its radar, however, is notoriously sensitive and often picks up false returns (caused by ground clutter) if trying to detect targets at an altitude of less than 60m (200ft). Although the RPK-2 can resist an enemy's passive electronic countermeasures, the radar has difficulty auto-tracking targets at ranges of less than 7km. This is due to the high angular speed of supersonic aircraft at close distances. Also, the crewmen often have to reset the ZSU's radar system because the cathode-ray power source cannot

A platoon of ZSU 23-4s on maneuver during a Soviet exercise. Soviet Army practice was that the ZSU 23-4 travelled alongside the armor and mechanized infantry formations. (US Army)

E

ZSU 23-4

The ZSU 23-4 was the most advanced air defense vehicle of its day. The 23-4 features a quad-mounted 23mm 2A7 gun system. These auto cannons fire a standard, steel-cased 23x152mm bullet at a muzzle velocity of 970m per second and have a combined rate of fire of 3,400–4,000 rounds per minute. Its most prominent feature, however, is the RPK-2 "Tobol" radar-tracking system.

The radar is a J-band device and can effectively track moving targets up to 10km away, providing good range data for both high-performance and subsonic aircraft. For acquiring targets at lower altitudes, the radar is accompanied by a high-powered optics system providing 2x and 6x magnification. However, the RPK-2 is not without its drawbacks. This radar is notoriously sensitive and often picks up false returns caused by ground clutter. The crewmen have also complained about the frequent need to reset the ZSU's radar because the cathode-ray power source cannot effectively track high-performance aircraft for long periods of time. Upgraded versions of the 23-4, however, include a solid-state electronic system for the radar, which has greatly improved its operability.

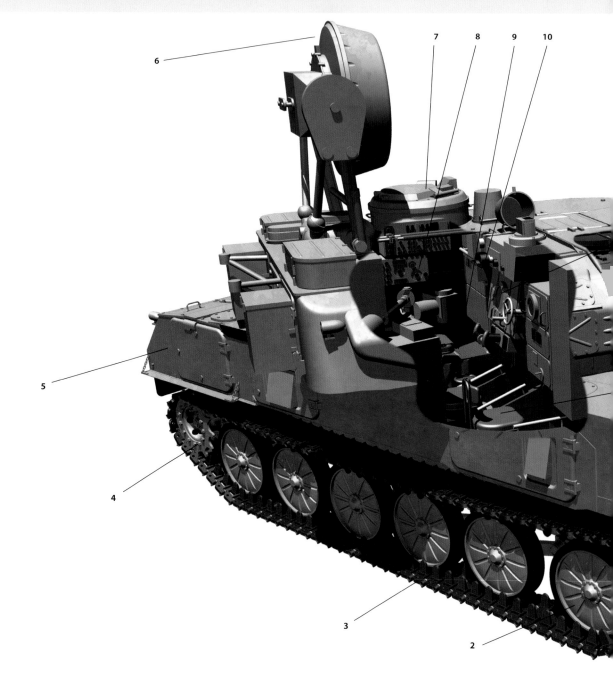

KEY

1. Front drive sprocket
2. Roadwheel
3. Track
4. Rear drive sprocket
5. Stowage
6. Tobol radar system
7. Commander's instrument hatch
8. Commander's panel
9. Gunner's station
10. Aiming reticle
11. Angular targeting interface
12. Quad 23mm autocannon
13. Ammunition feed
14. Radar operator's station
15. Driver's seat
16. Driver's instrument panel
17. Circuit box
18. Fire suppression system
19. Headlamp
20. Tow aperture

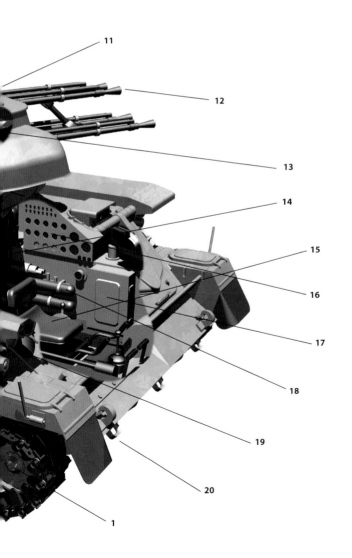

This pictorial, published by the US Department of Defense in 1981, illustrates seven tactical weapons that were considered the biggest threats to United States forces. Chief among them was the ZSU 23-4, as the Americans had no comparable air defense vehicle at the time. (US Department of Defense)

effectively track high-performance aircraft for long periods of time. In addition, the absence of any laser range finder means that only a highly skilled ZSU commander and gunner could destroy a target.

Since the late 1990s, however, the Russian Defense Ministry has made deliberate efforts to improve the ZSU 23-4's fire control and tracking system. The ZSU's most recent modernization program upgraded the radar from a cathode-ray device to a solid-state electronics system. This greatly improved the operability and endurance of the radar system, making it easier to acquire high-performance, high-angle targets. This radar also uses digital signal processing and has been finely calibrated to avoid the false readings that are frequently caused by ground clutter. The newer digital computing system has greatly enhanced the ZSU's tracking and gun laying capabilities. As a result, Russia's upgraded variants of the ZSU 23-4 can achieve target kill probabilities of 0.07 to 0.12 against aerial targets flying at steep angles. The upgraded fire control system also allows the ZSU to conserve its ammunition. Throughout its field trials, the upgraded ZSU expended only 300–600 rounds for each target destroyed. Previous versions of the ZSU had expended nearly 3,300–5,700 rounds under the same conditions.

The information gathered by the ZSU's radar system is sent directly to the onboard computer, which uses a three-plane stabilization system to calculate the intercept point and lead angle. With this information, the guns are automatically positioned on the target. This advanced targeting system allows the 23-4 to fire on the move at speeds of up to 25km/h (15.5mph) and on an inclined surface of up to 10 degrees.

The ZSU 23-4 has three firing modes: Radar Only, Radar Plus Optics, and Optics Only. In Radar Only mode, the radar provides the target azimuth, elevation, and range – enabling the onboard computer to calculate a full gunnery solution in a matter of seconds. A ZSU's gunner frequently uses this mode when engaging high-performance aircraft at lower altitudes. Radar

Plus Optics only provides the range to the target. The crewmen, using either the 2x or the 6x magnification, use the optical sight to track azimuth and elevation. In this mode, however, the computer will still control the gun placement. The ZSU 23-4 uses this mode against slower-moving, low-altitude targets or ground targets. For the Radar Only and Radar Plus Optics mode, the target computer's memory can assist with tracking targets that fly behind any type of obstruction. For example, if the radar loses sight of an enemy helicopter when it flies behind a hill, the computer's memory will provide the radar with 8–10 seconds of steering data based on the helicopter's projected flight path. That way, when the helicopter emerges from behind the hill, the ZSU's radar is waiting to pick up the aircraft without having to reacquire the target. In Optics Only mode, the operator inputs the data based on his own estimates or the observations of another gun crew. In this mode, the guns are pointed and tracked manually.

Despite its formidable firepower and advanced radar system, the ZSU 23-4 was not without its drawbacks. Earlier versions of the SPAAG had a primitive air-cooling system for the 2A7 guns. This design often led to having a "runaway gun." After long periods of firing, the residual heat of the barrel would cause the last chambered round to ignite and discharge from the gun. As with any automatic-fire weapon, once a round left the chamber, the feed belt would instantly load another round, which would also ignite and fire from the gun. Often, this chain reaction would continue until the entire belt of 23mm ammunition had been expended. Understandably, this made the vehicle highly dangerous to nearby ground troops. Although the gun's air-cooled system was later replaced with a water-cooled variant, Soviet field commanders generally kept the ZSU 23-4 away from their dismounted troops.

Once the ZSU 23-4 was fielded to the Soviet Army, the vehicle was integrated into the standard air defense battalions. Each of the battalions had two SPAAG platoons, with four ZSUs per platoon. These platoons were then parceled out to the motorized rifle regiments and tank units to provide short-range air defense. By the end of the 1960s, one SPAAG platoon in each

Crewmen prepare to load the 23mm shells into the ZSU's quad-mounted gun. The 23mm auto cannon had an impressive rate of fire, but earlier versions of the gun were highly susceptible to overheating and runaway firing. (Polish Defense Ministry)

An East German ZSU 23-4 at the Dresden Military History Museum. Like its predecessor, the ZSU 57-2, the newer ZSU was widely exported to Warsaw Pact countries. Following German reunification, most of the ZSUs within the Nationale Volksarmee were destroyed or given back to the Soviet Union. (Billy Hill)

battalion contained the ZSU 23-4, while the other operated the ZSU 57-2. By 1972, however, the last of the older 57-2s had been retired or given to allied nations. In the 1970s, the Soviet Union began reorganizing its tactical air defense formations. Under the new organizational scheme, one air defense battery would be placed within the motorized infantry and tank regiments. Each battery had two platoons: one platoon of four ZSUs and one platoon of four 9K31 Strela-1 missile launchers. The Strela-1 was a short-range, infrared, surface-to-air missile system mounted atop a BRDM-2 wheeled reconnaissance vehicle. Its four-tubed missiles could reach speeds of Mach 1.8 and hit targets at altitudes of 4,000m. The ZSU 23-4 operated alongside the Strela-1 until the missile system was replaced by the newer 9K35 Strela-10.

Soviet practice was to mark the ZSU 23-4's firing positions near the forward edge of the battle area, but usually a few hundred meters behind the front echelon. This protected the lightly-armored ZSU from enemy tanks and anti-tank missiles, both of which could destroy a ZSU with one hit. During offensive operations, the ZSU would normally stay 500m behind the lead tanks. For those ZSUs defending the line, the separation distances were much greater (up to 1,000m), because defense forces require more time to maneuver and regress against an attacking enemy. During its first decade of service, the ZSU operated autonomously within its parent formations – there was no target marking from air defense headquarters. However, with the development of the PPRU (a mobile, early-warning radar system built on a MT-LBu chassis) in the late 1970s, the ZSU 23-4 and its companion 9K31 and 9K35 missile platforms could have a local target marker to facilitate their antiaircraft engagements.

Like its predecessor, the ZSU 57-2, the Shilka also found a great export market among Soviet allies. Members of the Warsaw Pact – such as East Germany, Poland, Hungary, and Bulgaria – were among the first to receive the exported ZSU 23-4. As the East German *Nationale Volksarmee* began to retire its fleet of 57-2s, it imported more than 100 ZSU 23-4s as replacements. Following German reunification, most of the frontline ZSU 23-4s were either destroyed or given back to the Soviet Union. However, a handful of East German ZSUs were acquired by Thales Nederland, a subsidiary of the defense company Thales Group. Thales intended to launch its own modernization program for the vehicle and integrate it into the Dutch armed forces. A prototype was completed in 1998, but the Royal Netherlands Army did not adopt the vehicle.

Polish ground forces received 150 Shilkas before the end of the 1970s, many of which remain in service today. Poland began its own modernization program for the ZSU 23-4 in 1998. Named the ZSU 23-4MP Biala, the Polish variant includes a digital targeting system and four PZR Grom missiles as a secondary armament. The PZR Grom was a shoulder-fired surface-to-air missile, packing a 72mm warhead. The Grom was later adapted to fit on to the ZSU 23-4MP and other Polish air defense systems, such as the AZM POPRAD and the ZUR-23-2KG. Aside from the improved targeting system and the Grom missiles, the Polish variant of the ZSU remains virtually unchanged from its Soviet forebear. Hungary and Bulgaria still operate a handful of ZSU 23-4s within their formations. Cuba, Vietnam, Angola, and North Korea were among the other Soviet allies to receive ZSU 23-4s and each recipient still operates limited quantities of the vehicle.

Following independence from the Soviet Union, Ukraine used its inherited Shilkas to begin its own modernization program for the vehicle in the late 1990s. The Ukrainian variant, known as the Donets, was designed by the Malyshev Tank Factory in Kharkov. Although it retains the original turret from the legacy ZSU 23-4, the Donets is built on a T-80 tank hull. The Ukrainians also upgraded the Donets's armament to include the 9K35 Strela-10 surface-to-air missile system. Also known by its NATO reporting name, SA-13 Gopher, the Strela-10 is a short-range, optical/infrared guided missile that was first developed in the late 1960s. Traditionally, the missile system had been mounted atop the Soviet MT-LB, a tracked amphibious assault vehicle. Though relatively small, the Strela-10 has a 3.5kg warhead and can reach speeds of up to Mach 2. After the breakup of the Soviet Union, many of the Strela-10 missiles were subsequently absorbed by Ukrainian Ground Forces and refitted into various air defense platforms. Like Russia, Ukraine has also found an export market for its ZSU 23-4; as of 2014, the Indian Army has agreed to purchase 138 of the Ukrainian ZSU Donets.

A Polish crew prepares a ZSU 23-4 for action. The vehicle's RPK-2 radar is a J-band device that can detect aircraft from up to 20km away. (Polish Defense Ministry)

The ZSU 23-4 on parade in 1984. Despite the ZSU's impressive radar and target-tracking system, the vehicle was not without its drawbacks. The RPK-2 radar, for instance, often picked up false returns caused by ground clutter. Additionally, because of its cathode ray power source, the RPK-2 frequently had to be reset. (US Department of Defense)

However, India already had the base models from many ZSU 23-4s, as about 100 Shilkas were delivered from the Soviet Union during the 1970s. Before bidding on the ZSU 23-4 Donets, the Indian Army had begun its own modernization program for their existing fleet of ZSUs. Launched as a joint venture between Bharat Electronics and Israel Aerospace Industries, the Indians' ZSU modernization package featured a new Caterpillar 359bhp power plant, an improved fire control system, and jamming-resistant, solid-state radar.

Since the fall of the Soviet Union, Russia has also put forward its own modernization program for the legacy SPAAG. These upgrades, which began in the late 1990s, included packages for both the armament and the chassis. The enhanced Shilka features a crew air-conditioning system and a reduced thermal signature to make it less visible to an enemy's infrared optics. The transmission was also upgraded with new hydrostatic steering. This increased the vehicle's speed to more than 60km/h and enhanced both its mobility and reparability. In addition to the 23mm guns, this enhanced ZSU features two 9K38 Igla (SA-18 Grouse) missiles mounted on either side of the turret. These missiles, like the Polish Grom, contain a 72mm warhead and can reach top speeds of Mach 2. The installation of these Igla missiles increases the ZSU 23-4's effective kill range to 5.2km.

Two ZSU 23-4s on maneuver, 1982. (US Department of Defense)

The ZSU 23-4 first saw combat during the Arab–Israeli wars of the 1960s and 1970s. Egypt and Syria used several ZSU 57-2s and ZSU 23-4s during their bids to overwhelm the Israeli Defense Forces. However, following their defeat in the Six Day

War, Egypt and Syria approached the Soviet Union to replace the weapons that had been lost during the fight. Prior to 1967, most of the equipment that the Arabs had received from the Soviet Union was older and slightly worn out. They were reliable weapons but, like the ZSU 57-2, most had become obsolete. This time, however, the Syrian and Egyptian emissaries struck a better deal – they would only accept the most advanced military equipment. To that end, both nations received their first complement of ZSU 23-4s.

By 1972, Egypt had integrated the ZSU 23-4 into its frontline air defense units and their crews were being trained by Soviet personnel. During the Six Day War of 1967, Egyptian ground forces had suffered heavy losses at the hands of the Israeli Air Force. The Egyptians intended to deploy their armored forces covered by a protective umbrella of antiaircraft fire from the ZSU 23-4, arrayed in serried ranks similar to Soviet practice. When Egyptian forces crossed the Suez Canal on October 6, 1973 – Yom Kippur, the Day of Atonement – Israeli aircraft were promptly dispatched to halt the incursion. Unfortunately, as they neared the Canal, Israeli fighters soon ran into Egyptian air defenders. On that first day of combat, Israeli squadrons lost at least ten aircraft.

A captured ZSU 23-4 employed by marines of the 3rd Amphibious Armored Vehicle Battalion, 1st Marine Division at Camp Pendleton, California, 1997. These marines were members of the opposing force, commonly called "OPFOR," during Exercise Kernel Blitz 97, a major amphibious exercise involving more than 12,000 US servicemen. (US Marine Corps)

Farther north, in the Golan Heights, Syrian ZSUs formed part of the Arab air defenses that downed more than 30 Israeli aircraft. However, the Syrian air defense batteries were not as successful as their Egyptian counterparts – the Israelis soon launched a devastating anti-air defense campaign that reduced the Syrian gun and missile crews by nearly 50 percent. Since the ZSUs and their missile-launching counterparts had no early warning radar capabilities, the Israeli planes could fly within striking distance of the Syrian formations and achieve total surprise. In light of their heavy losses, the Syrians withdrew most of their air defense assets, including the ZSU 57-2s and ZSU 23-4s, to protect the perimeter around Damascus.

Meanwhile, as the Egyptians crossed the Suez Canal, their frontline formations outran the cover of their own air defenses. Thus, the Israeli Air Force was able to provide unhindered air support as their ground forces pushed the Egyptians back towards the Canal. Now within striking distance of the air defense barrier, the Israelis decided to attack the enemy SPAAG and missile positions where they were most vulnerable: on the ground. Indeed, by October 17, Israeli tanks had made short order of the Egyptian air defense sites. One week later, on October 25, a final ceasefire was negotiated and the Yom Kippur War came to an end. By the end of the conflict, however, the Israeli Air Force determined that most of their losses had been at the hands of the ZSU 23-4 – incurred as the Israeli planes attempted to dive below the engagement ranges of the surface-to-air missiles. In 2011, during the opening months of the Syrian Civil War, Syrian ZSU 23-4s deployed to the streets of Aleppo and Damascus where they engaged rebel forces operating from windows and rooftops.

Nearly a decade after war with Syria and Egypt, Israel cast its eyes towards Iraq. Although the two countries shared no land borders, Iraq was one of Israel's enemies in the Arab world. Iraqi leader Saddam Hussein had made several threats against Israel, promising to use "weapons of mass destruction." Fearful that the Osirak reactor (located near Baghdad) would be used to that end, the Israeli Air Force launched a preemptive strike against Osirak on June 7, 1981. At the time, Iraq had an impressive arsenal of ZSU 23-4s and, because of the ongoing war with Iran, the Iraqi Army had increased its air defenses around Baghdad and the nearby Osirak reactor. The air defenses around Osirak proper were especially formidable: ranks of ZSU 57-2 and ZSU 23-4 SPAAGs reinforced with 2K12 Kub missile launchers. Despite these defenses, however, the Israeli F-16s were able to achieve total surprise as they dropped chaff to confuse the ZSU's radar-guided guns.[3] The chaff had a similar effect on the 2K12 launchers – some of the Iraqi crews fired their missiles, but none of them locked on to their targets. The invading F-16s thus dropped their ordnance on to the reactor, rendering its nuclear capabilities useless. In the ensuing confusion, the ZSUs defending Baghdad began firing into the sky, but it was too late – the Israeli jets had already cleared the airspace and were heading back home. The following year, during Operation *Peace for Galilee* (the brief war with Lebanon), Syrian ZSU 23-4s were back in action against the Israeli Air Force. This time, however, the Syrian ZSU crews managed to shoot down six Israeli aircraft.

Yemen took receipt of its ZSU 23-4s in the 1970s when it was also using American air defense vehicles such as the Hawk and M163 VADS. Col Valentin Nestereko, a Soviet air defense advisor to the Yemeni forces, recalled that the Yemenis were decidedly pro-American when it came to selecting military equipment. However, according to Nestereko, when they received their first ZSU 23-4 and tested its tracking radar, the Yemeni troops wholeheartedly embraced the ZSU 23-4. Today, about 30 ZSUs remain in service in Yemen.

The ZSU 23-4 was a late arrival to the Vietnam War. As the People's Army of Vietnam began to phase out its own ZSU 57-2s, they introduced the newer 23-4 in its place. The vehicle was used in the Ho Chi Minh Campaign of 1975 – the final conquest of South Vietnam – where it served with the 237th Antiaircraft Artillery Regiment. By this time, however, American air combat had ended and the US had withdrawn from Southeast Asia.

---

3   Chaff refers to a radar countermeasure whereby an aircraft deploys a small cloud of metal shards to disrupt the enemy's tracking capabilities.

In 1979, the Soviet Union sent troops into Afghanistan to facilitate the Afghan counterinsurgency against the mujahideen. However, the Soviet military and its tactical doctrines were ill-equipped to handle the complexities of asymmetric warfare. After nine bloody years, and no appreciable gains against the mujahideen, the Soviets quit the field and withdrew from Afghanistan. During the Soviet intervention, however, the ZSU 23-4 found its most prominent role. In the mountainous valleys of Afghanistan, the ZSU was stripped of its air defense capabilities and converted into a ground support and convoy escort vehicle. The

An aging M551 Sheridan vehicle has been modified visually to represent a ZSU 23-4, seen here at the National Training Center in Fort Irwin, California. Vehicles of this type, called "VISMODs," are a common sight at Fort Irwin as the resident opposing force strives to create a realistic Soviet-based enemy. (US Army)

resulting variant, the ZSU 23-4M2 Afghanskii, had its radar system removed and replaced with an enhanced night-vision device. This Afghanskii variant also had an increased ammunition capacity of 4,000 rounds.

Protecting convoys was a top priority for Soviet commanders in Afghanistan. One of the mujahideen's preferred tactics was to attack convoys along the main supply routes through the mountainous passes. Placing rocket-propelled grenade (RPG) teams along the cliffs and escarpments, the mujahideen would triangulate their attacks on Soviet columns and fire on them from heights that were beyond the normal elevation range of an armored vehicle's main gun. With the integration of the ZSU into the convoy security teams, the Soviets now had an effective means of dealing with the high-ground snipers. After a few years of war, Soviet columns that were accompanied by the ZSU 23-4 were rarely attacked by Afghan rebels.

But even the ZSU's resilience could not save convoys from annihilation. Often, the Soviets' field losses were due to their poor command decisions and top-heavy leadership that discouraged initiative. For instance, on December 2, 1981, the 2nd Reconnaissance Company (garrisoned in Maimana, Faryab

A ZSU 23-4 in service with the Afghan National Army. The USSR sent several ZSUs to the Afghan national forces before and during the Soviet intervention against the mujahideen. After the Soviet withdrawal, many ZSUs fell into the hands of the Taliban. Since the US-led invasion of 2001, the Afghan National Army has been re-equipped with and re-trained on many of its Soviet-era weapons. (RFI)

A ZSU 23-4 fires its auto cannons at a leveled position. Despite its intended role as an air defense weapon, the ZSU 23-4 has seen more combat as a direct-fire ground support vehicle, particularly during the wars in Afghanistan and Chechnya. (Polish Defense Ministry)

Province) escorted a convoy of 120 vehicles from Andkhoy. The company reinforcements included a sapper squad and a single ZSU 23-4. The reconnaissance group commander, Lt Col A.A. Agzamov, recalled the disaster that ensued when the convoy and its escort vehicles reached the village of Daulatabad:

> [The forward patrol] reported back that the village was deserted. This report put us on our guard, and the convoy commander ordered us to increase our observation. When the lead vehicles of the convoy began to exit Daulatabad village, the enemy opened fire with a grenade launcher and destroyed a BMP-2KSH and a fuel tanker. A fire broke out and the vehicles immediately behind the conflagration were stuck in narrow streets.
>
> The enemy opened up with small arms fire. Two more BMPs were knocked out and, as a result, the convoy was split into three sections. We returned fire, but it was not controlled or directed. The convoy commander lost control over his sub-units since his communications were gone. Individual vehicles independently tried to break out of the kill zones. The [forward air controller] called in helicopter gunships and directed their fires. The helicopters began gun runs on the enemy in the village. In the meantime, the trail platoon received the mission to sweep the western part of the village. The dismounted troopers moved under the cover of BMP and helicopter fire to carry out their mission.

ZSU 23-4 IN AFGHANISTAN

The ZSU 23-4 fires at enemy snipers along a mountain pass in Afghanistan. Although an air defense weapon by design, the ZSU 23-4 was widely used as a convoy security vehicle during the Soviet War in Afghanistan. The ZSU's quad-mounted guns were highly effective when engaging snipers who hid atop mountainsides, cliffs, and other high places that were beyond the reach of T-72 or BMP main guns. The Soviets even developed a modified version of the 23-4 for combat in Afghanistan. Known as the ZSU 23-4M2 "Afghanskii," this modified ZSU had an added night-sight, increased ammunition capacity, and its radar was removed as the mujahideen had no air assets.

The ZSU 23-4MP "Biala," a Polish upgrade of the legacy SPAAG, features the Grom surface-to-air missile system, and a new digital aiming system. Many of the former communist nations that imported the ZSU have begun their own modernization programs for the vehicle. (Polish Defense Ministry)

The enemy withdrew when faced with this decisive action. In the course of this three-hour battle, four of our soldiers were killed (all drivers), six were wounded, three BMPs were destroyed and five trucks were burned up.[4]

Although the attendant ZSU 23-4 survived this episode, the ill-fated convoy mission highlighted the fundamental problems in the Soviets' tactical bearings. For this operation, there had been no adequate preparation for combat and the vehicle commanders had not been briefed on probable enemy locations or courses of action. Furthermore, the convoy commander would have done well to dismount his troops and sweep the village to intercept any ambush teams. Only when close air support arrived did the convoy commander regain the upper hand.

Despite its effectiveness against the Afghan insurgents, the ZSU remained vulnerable to large caliber weapons – generally anything greater than a 7.62mm weapon. The mujahideen were able to destroy a few ZSUs by using "broken rhythm" attacks. Under this method, the Afghan rebels would launch a diversionary RPG attack to draw the ZSU's attention while another RPG team would engage the ZSU from a different position. The success of these broken rhythm attacks varied depending on how well each vehicle in the convoy maintained its assigned sectors of fire. In other instances, the vehicles were caught within the narrow confines of a village's streets and simply overwhelmed by multi-directional fire. Such losses occurred in May 1980, when a Soviet motorized rifle regiment was operating in the area surrounding the Salang pass between Kabul and Termez. To minimize the mujahideen's activity along the road, one of the regiment's motorized battalions was task-organized as a raiding detachment and supplemented with a platoon of ZSU 23-4s. After raiding a village in the Ghorband district – an operation that killed ten mujahideen fighters – the detachment moved northward to Bamian. On the road to Bamian, however, the detachment fell into an enemy ambush, and the air defense platoon lost a ZSU 23-4 in the process.

4 Grau, Lester. *The Bear Went Over the Mountain*, p.111.

Still, the ZSU 23-4 could be highly effective when rendering direct fire support to dismounted infantry. In March 1986, a platoon of ZSU 23-4s took part in an assault on an enemy weapons cache in the Xadigar Canyon, Kandahar. The assault force was impressive: two motorized rifle battalions, a D-30 howitzer battalion, a Spetsnaz detachment, two attack helicopter squadrons (Mi-8 and Mi-24, respectively), and a squadron of Su-25 aircraft in addition to the ZSU platoon. According to Lt Col S.Y. Pyatakov, the ZSU 23-4s gave direct fire support to the dismounted Spetsnaz troops that air assaulted into the canyon. With the ZSUs behind them, the Spetsnaz swept the canyon for any mujahideen fighters or equipment that remained following the airstrike and D-30 artillery preparations. The operation was successful in that the aerial bombardment had suppressed and/or destroyed most of the enemy's equipment before the main assault force arrived. By the end of the assault, the Soviets had killed 20 mujahideen fighters and, according to Pyatakov's report, had captured "a large amount of ammunition, documents and combat equipment of the guerrilla groups. There were no Soviet casualties."

The ZSU 23-4 repeated its role as a counterinsurgency weapon during the Chechen Wars of the 1990s. Following the collapse of the Soviet Union, Chechnya, a remote province in southern Russia, sought its independence. The ensuing struggle between Chechen nationalists and President Boris Yeltsin led to Russian military intervention in December 1994. The First Chechen War, as it became known, was a disaster for the Russian Ground Forces. Their first foray into the Chechen capital, Grozny, ended with more than 2,000 casualties and 150 armored vehicles lost. One of the problems faced by the Russian Ground Forces was that the main guns onboard the T-72s and BMPs could neither elevate high enough to engage insurgents on rooftops nor depress low enough to engage the insurgents firing from basement windows. Withdrawing from the city to regroup, the Russians re-evaluated their tactics and divided their forces into battalion-sized assault elements. This time, dismounted infantry would take the lead while T-72s and BMPs would assume a supporting role. Since the assault forces needed a

An upgraded ZSU 23-4M4 at the Russian Arms Expo in 2013. This modernized variant features an additional 9k38 surface-to-air missile system and an improved tracking radar. (Vitaly V. Kuzmin)

weapon capable of defeating the insurgents on rooftops and in basements, they called in the ZSU 23-4. The Russians' second foray into the city proved less costly than the first – their infantry and ZSU teams succeeded in suppressing many of the insurgents, but the Chechen affair proved too costly for Russian Ground Forces, and they subsequently withdrew in the fall of 1996. It was not until the Second Chechen War in 1999–2000 when Russian forces finally re-established federal control in the area.

Iraq, after fighting the Iranians to a draw during their eight-year war, put the ZSU 23-4 on the frontlines once again in the Gulf War of 1991. Iraqi forces, still under the command of Saddam Hussein, invaded Kuwait in August 1990, prompting UN condemnation and a US-led military effort to expel the Iraqis from the emirate. During the air campaign against Iraq, several ZSU 23-4s were destroyed by missile and guided bomb attacks. At the start of the ground war, the ZSUs that remained functional were arrayed in textbook formation to meet US armor in what Saddam Hussein predicted would be the "Mother of all Battles." However, many of these ZSUs were easily destroyed by M1 Abrams tanks and M2/M3 Bradley Fighting Vehicles during the 100-hour ground campaign. Despite the Iraqi Army's humiliating defeat in the Gulf War, the ZSU 23-4 remained in service over the next decade. In the opening days of Operation *Iraqi Freedom*, an American Unmanned Aerial Vehicle (UAV, or "drones" as they are commonly known) successfully destroyed a ZSU 23-4 with its Hellfire missile near the town of Al Ammarah on March 22, 2003. This target was the first confirmed kill by a UAV in Iraq.

A destroyed Iraqi ZSU 23-4 in the Euphrates River Valley following Operation *Desert Storm*, March 4, 1991. The oil drum on the back of the vehicle was an Iraqi modification featured on many of its heavier vehicles, including the T-72 tank. (US Department of Defense)

9K22 TUNGUSKA (SA-19 GRISON)

The development of the 9K22 Tunguska marked a radical step forward in the evolution of Soviet tactical air defense. A tracked, self-propelled antiaircraft vehicle, the Tunguska was the first Soviet vehicle to feature both antiaircraft guns and an integrated missile system. Preliminary design work on the Tunguska began on June 8, 1970 at the KBP Instrument Design Bureau in Tula. The Soviet Defense Ministry selected the KBP plant for its successful track record in developing various auto cannons and anti-tank missiles for the Red Army. The project began as a response to the observed shortcomings of the ZSU 23-4. Although the ZSU was a reliable vehicle and a potent air defense weapon, its firing range was still relatively short and it had no built-in early warning systems. As mentioned previously, the PPRU had been developed as a mobile radar system to provide an early warning and target-marking capability to the armored formations. However, Soviet defense planners ultimately sought to combine these capabilities into a single platform. The Tunguska was also created to counter the new class of ground attack aircraft (namely the A-10 Thunderbolt and the AH-64 Apache), which were purportedly resistant to 23mm gunfire.

During the initial design phase, KBP determined that a 30mm cannon would require at least three times fewer shells than the 23mm gun onboard the ZSU. A comparison test fire revealed that the 30mm cannon had a greater kill probability than the 23mm when targeting level aircraft flying at 300m/sec. The larger caliber weapon also increased the maximum engagement altitude from 2,000m to 4,000m. As the ZSU's gun system and the existing missile platforms had similar fire control systems, Soviet planners decided that the Tunguska would be a combined gun-missile system.

Preliminary designs of the Tunguska were completed in 1973 and the first round of prototypes delivered in 1976. Further development of the vehicle slowed, however, in 1975 after the 9K33 Osa missile system was fielded to Soviet ground forces. The 9K33 (NATO reporting name: SA-8 Gecko) was a

A 1986 artist's impression of the next-generation Soviet air defense vehicle. When this picture was drawn, US intelligence was convinced that the Soviet Union was working on a vehicle that combined antiaircraft guns with surface-to-air missiles. As it turned out, this new air defense system was the 9K22 Tunguska, and looked remarkably similar to this drawing. (US Department of Defense)

The 9K22 Tunguska marked a new chapter in Soviet air defense weapons. Like the ZSU 23-4 Shilka, the Tunguska was also named after a river in Siberia. Featuring two 30mm antiaircraft guns and eight 9M311 missiles, the 9K22 was also known by its NATO reporting name: SA-19 Grison. (US Department of Defense)

wheeled, 6x6, amphibious missile platform featuring six surface-to-air missiles with greater performance metrics than the missiles that were currently slated for the Tunguska. After considerable debate, however, Soviet defense ministers decided that a pure missile system would be less effective against attack helicopters. Aircraft of this type typically flew below the normal scanning levels of an early warning radar – and the Israelis had used this technique with considerable success during the Yom Kippur War. The Soviet design teams also recognized that the reaction time for a gun system was 8–10 seconds while even the best missile system had an engagement time of nearly 30 seconds. Thus, development of the 9K22 went forward once again. Field trials began in September 1980 and continued until December 1981. The Tunguska was officially accepted into service on September 9, 1982 and has had several designations throughout its service life. As the vehicle carried both missiles and antiaircraft guns, the Soviet Army dropped the ZSU designation and instead named it 9K22, although it was also listed in the Soviet registry as 2K22 and 2S6. On the other side of the Iron Curtain, its NATO reporting name was SA-19 Grison. The Tunguska entered limited service in 1984 and was fully integrated into the Soviet Army by 1990.

The base model Tunguska was built atop the GM-352 chassis, developed by the Minsk Tractor Works. The GM-352 was one of many tracked chassis built under the common GM (General Machine) series. Other GM models included the GM-5955 built for the 9K330 Tor missile system and the GM-5975 for the improved Tunguska M1. Unlike most GM-based vehicles, however, the Tunguska featured six road wheels instead of seven. Each set of tracks has hydro-pneumatic suspension to facilitate maneuvers on rough

A 9K22 Tunguska during the 2008 Moscow Victory Day Parade. Since its introduction in the 1980s, the Tunguska has only seen action during the 2008 South Ossetia War and the 2014 Crimean Crisis. (Theman Naziro)

terrain. Also integrated into the chassis was a Nuclear, Biological, and Chemical (NBC) protection system. The NBC system provides over-pressurization of the air inside the vehicle to prevent or minimize cross-contamination of radioactive or chemical elements. Furthermore, it allows the crew to conduct extended operations in a contaminated environment. The over-pressurization module is accompanied by an automatic fire-suppression system that activates the built-in fire extinguishers in both the driver and command crew compartments.

The Tunguska's power plant is a 12-cylinder, turbo-charged, liquid-cooled, four-stroke, fuel-injected, B-46-6 MC diesel engine. This 12-cylinder power plant can produce up to 840 horsepower and has been adapted to operate in the harsh extremes of the Russian climate. For example, the engine can start in temperatures as low as -5 degrees Celsius (23 degrees Fahrenheit) and can maintain operability in temperatures as low as -50 degrees Celsius (-58 degrees Fahrenheit). The Tunguska can also negotiate slopes of an incline of 35 degrees and can ford streams at a depth of 1m.

Like its predecessor SPAAGs, the Tunguska's turret rotates 360 degrees and houses most of the vehicle's crew (commander, gunner, and radar operator). The dual 30mm guns, known as the 2A38, are manufactured by the Tulamashzavod Company. Each of these auto cannons alternately fire 1,950–2,500 rounds per minute. Bullets fired from the 2A38 have a muzzle velocity of 960m/sec and can fire bursts of 80–200 rounds depending on the target type. These bullets are fed through a chain link and can reach altitudes of up to 3km. Although sources vary on the 2A38's elevation ranges, it generally falls within +85 degrees to -10 degrees, thus giving it ample room to engage both aerial and ground targets. The auto cannon aboard the 9K22 can be fired in one of two modes: Radar or Optical. In Radar Mode, as the name implies, the radar tracking system adjusts the gun lay and the rate of traverse. In Optical Mode, the gunner tracks the target through the Tunguska's 1A29 stabilized sight. In this mode, the radar only provides the range to target.

The radar system on the 9K22 combines two separate radars. The first is a parabolic antenna located on the top-rear of the turret. This E-band device (operating within a frequency range of 2–3 GHz) is the primary means for target acquisition. The second radar is a front-mounted J-band monopulse system that facilitates target tracking as the target moves along its flight path. The monopulse configuration sends out dual radar beams and, when the beams are reflected back to the vehicle, the onboard computer compares the beams to determine which of them is the strongest. This allows the system to correct for any changes in the target's speed or elevation.

The Tunguska's radar and fire control system has five operating modes:

1: Automated radar tracking.
2: Manual electro-optic tracking with range data provided by radar.
3: Inertial tracking.
4: Manual electro-optic tracking with range estimation provided by radar.
5: Ground engagement.

In the automated tracking mode, tracking data is fed directly to the computer once the radar has locked on to a target. At this point, the 1A29 sight can track the line of sight to the target or be used independently to track other targets in the area. Using the data calculated from the fire computer, the Tunguska's weapons are automatically laid on the target and the crew selects the appropriate weapon for engagement. When using the 9M311 missiles, the gunner must track the target optically for the duration of the engagement. With this automated tracking mode, the Tunguska is said to have a kill probability of 65 percent. Electro-optic and inertial tracking modes are intended for use during degraded conditions. However, these modes drastically increase the crew's engagement time, are less accurate when engaging targets, and must be performed while the vehicle is stationary. During the ground engagement mode, the radar system is turned off and the gunner uses the optical sight. Using this method, the Tunguska automatically calculates the lead angle, and lays the gun in proportion to the gunner's movement.

The 9M311 system carries four missiles – two in each firing block mounted on either side of the vehicle. Both firing systems can be elevated and fired independently. The individual 9M311 rockets are powered by a solid propellant fuel and have an individual weight of 57kg, including a 9kg warhead. The warhead itself is a continuous-rod system, meaning that an even number of rods are welded together within the warhead to form a cylindrical pattern. When the missile detonates, it expels the rods into a large circular pattern designed to cut through the target. The welded rods within the 9M311's warhead measure 600mm in length and vary from 6 to 9mm in

G

9K22 TUNGUSKA

The 9K22 Tunguska (NATO reporting name: SA-19 Grison) is a tracked air defense vehicle that features anti-antiaircraft guns combined with surface-to-air missiles. The Tunguska's primary armament is its eight 9M-series missiles. To engage closer targets, the crew can switch to its twin 30mm 2A38M guns. The 9K22 entered service in 1982, but remained in development throughout most of the 1980s. Today the Tunguska is still used by the Russian Ground Forces (1), as well as the armies of Ukraine (2) and Belarus. Despite its lengthy service, however, the Tunguska has not seen much combat. In Eurasia, its only recorded deployments have been during the Russia–Georgia conflict of 2008 and the Crimean crisis of 2014.

1

2

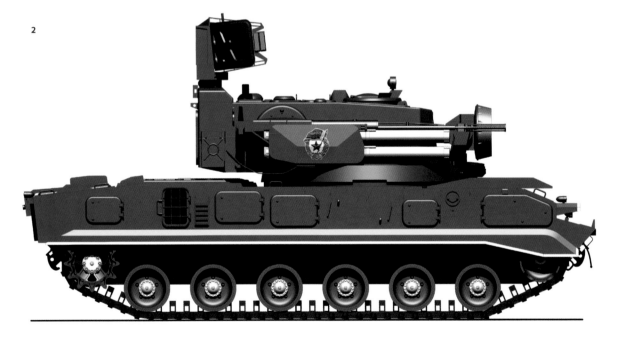

A Tunguska air defense vehicle in Ukrainian service. Like many of the Soviet-era SPAAGs, former communist republics such as Ukraine and Belarus absorbed the 9K22 into their new armies. Here, a 9K22 rolls past a crowd of onlookers at the 2008 Independence Day Parade in Kiev. (Vitaly Antonjuk)

diameter. Configured into a flower-like cross-section, these rods break into fragments weighing 2–3g on detonation.

When fired, the 9M311 missile has a two-stage deployment system. In the first stage, a four-fin booster pushes the rocket to a maximum velocity of 900m/sec (Mach 3). Once this booster falls away, the second stage begins, activating the primary missile featuring four fixed fins, each with steerable control surfaces. Here, the Tunguska's radar system guides the missile to its target. As the radar continues to track the target (including range, elevation, speed, and direction), the onboard fire computer will generate commands to adjust the missile's trajectory. The gunner must also track the missile to the target using the 1A29 optical sight in 8x magnification. Under this method, the gunner tracks the azimuth and elevation, but manually sets the linear range. Once the missile arrives within 5m of the target, the 9M311 will automatically trigger its proximity fuse. However, the 9M311 also has an impact fuse as a back-up measure. The 9M311 can engage aerial targets moving at speeds of up to 500m/sec and at linear ranges of up to 8,000m. During the initial launch sequence, the Tunguska must be stationary to avoid damaging the missile as it leaves the tube. After the missile is airborne, the launch tube is lowered to keep a clear line of sight to the target.

Since its introduction in the 1980s, the Tunguska has gone through two successive variants. The initial production model was known as the 2K22 (or 9K22M) and featured the base model 9M311 missiles. In 2003, the Russian Ground Forces fielded the Tunguska M1 model. The M1 featured an improved 9M311 missile with a range of 10km and an improved fire control system. This variant entered service on July 31, 2003.

Like its predecessors, the 9K22 provides a mobile and flexible air defense for armor and motorized regiments. According to Russian practice, the Tunguska has its own battery within the regimental air defense battalion. A 9K22 battery is composed of a headquarters section, transportation section, and three air defense platoons. Each platoon has two 9K22s, for a total of six per battery.

Another Tunguska stands in the queue for the 2008 Moscow Victory Day Parade. The radar system aboard the 9K22 combines a rear-mounted parabolic E-band device with a front-mounted monopulse J-band tracker. This radar system can track targets at altitudes of up to 3,500 meters. (Vitaly V. Kuzimn)

As the regiment's primary air defense weapon, the Tunguska performs a variety of missions. These include but are not limited to defending armored columns, setting air defense ambushes, and establishing a roving air defense patrol. When defending armored columns on a tactical march, the 9K22s are divided into pairs and placed at least 1,000–2,000m from one another to optimize their interlocking fields of fire. In the air defense ambush, two 9K22s take up static positions and only engage targets that come within a designated aerial sector. The Tunguska SPAAGs relocate after engaging a target or upon discovery by enemy forces. In the roving defense, instead of lying in wait for the enemy's aircraft, the Tunguska teams move to whatever area is most likely to fall under the enemy's air attack.

Unlike the ZSU 57-2 and ZSU 23-4, the 9K22 Tunguska has not seen a wide export market. Russian Ground Forces remain the primary operator of the vehicle, with more than 250 currently in service. After the breakup of the Soviet Union, the various successor states (including Ukraine and Belarus) retained the existing fleets of 9K22s that remained within their borders. The Indian Army, a long-standing customer for Soviet exports, continues to operate the 9K22 alongside the T-72, T-90, and BMP. During the past decade, the Tunguska has also been exported to Syria and Morocco.

TRENDS IN POST-SOVIET SPAAGS

Although the ZSU 57-2 has been withdrawn from Russian service (and from most of the former Warsaw Pact), it nevertheless remains operational with select militaries around the world. There can be little doubt that the ZSU 57-2 has been outclassed both as an air defense weapon and as a ground support gun. However, the militaries that still employ the vehicle have limited defense budgets and have thus far kept the older ZSU serviceable by fashioning their own parts for it. Taken together, these conditions will likely extend the ZSU 57-2's service life well into the 21st century.

Despite its age, the ZSU 23-4 continues to be a staple of modern air defense. The Russian Ground Forces, Russian Marines, and many armed forces throughout Eurasia still operate the vehicle in one of its many upgraded formats. Since the advent of vehicles such as the Tunguska (combining antiaircraft guns with surface-to-air missiles), countries such as Russia, Poland, and Ukraine have upgraded the ZSU 23-4 to include missile launchers and improved tracking systems. Also, because the ZSU 23-4 was produced in such great numbers, these modernization programs present the most cost-effective means of prolonging the ZSU's life until it can be fully replaced by the newer systems such as the Tunguska.

The tracked variant of the Pantsir-S1 integrated gun-missile air defense system. Also known by its NATO reporting name, SA-22 Greyhound, the Pantsir-S1 is a further development of the 9K22 and uses a phased-array radar system. (Vitaly V. Kuzmin)

The Tunguska itself, however, may be replaced by a newer air defense system known as the Pantsir-S1 (NATO reporting name: SA-22 Greyhound). Developed in 1990, the Pantsir-S1 did not fully enter service until the early 2000s. Like its 9K22 predecessor, the Pantsir-S1 features dual 30mm auto cannons and an integrated missile system. Unlike the Tunguska, however, the Pantsir-S1 carries the improved 57E6-series missile and is not limited to a tracked chassis. Indeed, the Pantsir-S1 is the most versatile air defense platform used by the Russian Ground Forces; its integrated gun-missile system can be placed atop a tracked GM-352 chassis or almost any variety of heavy-duty 8x8 trucks. There is also a naval version of the system that Russia plans to integrate on board its warships. Although the Pantsir-S1 will probably replace the ZSU 23-4s and 9K22s, the resilience of the latter two systems (and the viability of their modernization packages) indicates that they will remain in service for years to come.

The Pantsir-S1 mounted on an 8x8 KAMAZ truck. Development of the Pantsir began in 1990 and continued after the collapse of the Soviet Union. Following years of development, the air defense system finally debuted with Russian Ground Forces in 2003. (Vitaly V. Kuzmin)

FURTHER READING

Andrade, Dale. *America's Last Vietnam Battle: Halting Hanoi's 1972 Easter Offensive*. University Press of Kansas: Lawrence, 2001

Citino, Robert. *Blitzkrieg to Desert Storm: The Evolution of Operational Warfare*. University Press of Kansas: Lawrence, 2004

Cordesman, Anthony. *The Lessons and Non-Lessons of the Air and Missile Campaign over Kosovo*. Praeger: Westport, 2001

Crabtree, James. *On Air Defense*. Praeger: Westport, 1994

Gott, Kendall. *Breaking the Mold: Tanks in the Cities*. Combat Studies Institute Press: Fort Leavenworth, 2006

Grau, Lester. *The Bear Went Over the Mountain*. National Defense University Press: Washington DC, 1996

Grau, Lester. *The Soviet–Afghan War: How a Superpower Fought and Lost*. University Press of Kansas: Lawrence, 2002

Oliker, Olga. *Russia's Chechen Wars, 1994–2000: Lessons from Urban Combat*. RAND Corporation: New York, 2001

Tradoc Bulletin 4: Soviet ZSU 23-4: Capabilities and Countermeasures. Army Training and Doctrine Command: Fort Monroe, 1976

Zaloga, Steven J. *ZSU 23-4 Shilka and Soviet Air Defense Gun Vehicles*. Concord Publications: Hong Kong, 1992

INDEX

References to images and plates are in **bold**.

9K22 (Tunguska) 4, **43**, 46
  armament 39–40, **40**, 41, 42, 44
  chassis 40–41
  crew 41
  deployment 44–45
  development 39–40, **39**
  engine 41
  exports **44**, 45
  fire control system 42
  Nuclear, Biological, and Chemical (NBC) protection system 41
  radar system 42, 45
  turret 41
  variants 44
9K31 Strela-1 missile launchers 28
9K33 Osa missile system 39–40
9K35 Strela-10 surface-to-air missile system 28, 29
9K38 Igla (SA-18 Grouse) missiles 30
9M311 missile system 42, 44

Afghanistan 4, 33–34, **33**, **34**, **35**, 36–37
Agzamov, Lt Col A.A. 34
Algeria 16
ammunition 8, 10, 12, 20, **24**, **27**, 41
Angola 16, 29
antiaircraft guns 8
armament
  9K22 (Tunguska) 39–40, **40**, 41, 42, 44
  9M311 missile system 42, 44
  DshK machine guns 8
  Pantsir-S1 46
  PZR Grom missiles 29
  ZSU 23-4 20, 27, **27**, 30
  ZSU 37-2 19
  ZSU 57-2 9, **9**, 10, 13
  ZSU-37 **7**, 8
armor
  ZSU 23-4 19
  ZSU 57-2 12
attack helicopters 40

Balkan Wars 14, 18
Bamian, action at 36
Belarus 45
Bharat Electronics 30
Bulgaria 14, 16, 28, 29

Chechen Wars 4, 37–38
counterinsurgency 37–38
Croatia **17**
Croatian War of Independence 18
Cuba 16, 29

Daulatabad, action at 34, 36

East Germany 14, 28, **28**
Egypt 16, **17**, **22**, 30–31
elevation range 4, 10, 41
engines
  9K22 (Tunguska) 41
  ZSU 23-4 19
  ZSU 57-2 12
  ZSU-37 7
Ethiopia 16
Experimental Design Bureau No. 16 19
exports 4, 10, 13–14, **15**, 16–18, **17**, **28**, **44**, 45

Finland **17**
fire control 23, 26, 42
fire support 37
First Gulf War 18, 38
Flakpanzer Gephard 5–6, **6**

Grabin, Vasily 9
Grozny 37
GU-PVO 6

Hungary 14, 28, 29

India 30, 45
Iran 16, **32**
Iraq 16, 18, 32, **32**, 38
Israel 16, **17**, 30–2, 40
Israel Aerospace Industries 30
Israeli Air Force 31, 32

Jordan 16
Joseph Stalin Factory No. 92, Gorky 9

KAMAZ truck **46**
KBP Instrument Design Bureau 39
Korean War 5

M42 Duster **5**, 17
M163 Vulcan Air Defense System (VADS) 5, **6**, 20
M247 Sergeant York Division Air Defense (DIVAD) vehicle 5–6, **6**
Malyshev Tank Factory 29
Minsk Tractor Works 40
Morocco 45
Mozambique 16

NATO 5
Nestereko, Col Valentin 32
North Korea 16, 29
Nuclear, Biological, and Chemical (NBC) protection system 41

Ob'yekt 500 9
Operation *Barbarossa* 6
Operation *Desert Storm* 4, 32, 38, **38**
Operation *Iraqi Freedom* 38
Operation *Peace for Galilee* 32

Pantsir-S1 46, **46**
People's Army of Vietnam (PAVN) 4, **14**, **15**, 16, 16–18, 32
Poland 10, 14, **17**, 28, 29, 37, 46
Pyatakov, Lt Col S.Y. 37
PZR Grom missiles 29

radar systems 13, 23, **24–25**, 26, 29, **30**, 39, 42, **45**
rate of fire 10, 13, 19, 20, **27**
Red Army, mechanized forces 4
Resolution 426-11 19
Romania 14
Royal Netherlands Army 29
RPK-2 radar 23, **24–25**, 26, 29, **30**
Russia **32**, 46
Russo–Finnish War 6

Six Day War 4, 16, 30–31
South Ossetia War 4
Soviet Union, collapse of 4, 6, 37, 45
SU-76M field gun 8
Suez Canal 31
Syria 16, **17**, **22**, 30–31, 45

T-90 SPAAG 8
target acquisition 8, 12–13, 23, **24–25**, 26, 42
Thales Nederland 28–29
Tunguska. see 9K22
turret
  9K22 (Tunguska) 41
  ZSU 23-4 19
  ZSU 57-2 10, **10**
Type 63 17

Ukraine 29, **44**, 45, 46
US Army 5–6, **5**, **6**, 17, 20, **33**
US Department of Defense 4, **26**
US Marine Corps 31

Vietnam War 4, **5**, **14**, **15**, 16, 16–18, 32
VISMODs **33**

Warsaw Pact 10, 14, 28
Weinberger, Caspar 6
West Germany 5
Western developments 5, **5**

Xadigar Canyon, action at 37

Yemen 32
Yom Kippur War 4, **22**, 31, 40
Yugoslavia, former 14, 18

ZiS-3 field gun 9
ZSU 23-4 4, **5**, 18, **19**, **21**, **22**, **23**
  armament 20, **24–25**, 27, 30
  armor 19
  combat service **22**, 30–34, **32**, **33**, **34**, **35**, 36–39, **39**
  counterinsurgency role 37–38
  crew 19
  development 13, 18–20
  drawbacks 27, **27**, 30, 39
  engine 19
  exports 16, 28–30, **28**, **29**
  fire support role 37
  firing modes 26–27
  firing positions 28
  formations 27–28
  ground support and convoy escort vehicle 33–34, **34**, **35**, 36–37
  kill probability 22, 26
  maneuverability 19, 20
  modernization 26, 30, **38**, 46
  optics system 23
  production 20
  radar system 23, **24–25**, 26, 29, 30, **30**
  speed 19, 30
  tactical countermeasures 22–23
  turret 19
  US threat perception **26**
  vulnerability 36
ZSU 23-4 Donets 29, 30
ZSU 23-4MP Biala 29, **36**, **37**, **38**
ZSU 37-2 19, 20
ZSU 57-2 4, **5**, **11**, **12**, **13**, **22**, 28
  armament 9, **9**, 10, 13
  armor 12
  chassis 10
  combat service 16–18, 32
  crew 10
  development 9
  drawbacks 12–13
  engine 12
  exports 10, 13–14, **15**, 16–18, **17**
  maneuverability 12
  modernization 13
  range 12
  rate of fire 10
  service life 45
  turret **10**
ZSU-37 4, **7**, **8**
  armament **7**, 8
  crew 8
  development 8
  drawbacks 9
  engines **7**
  production 8
  speed 8